I0486395

*

*

GRAVITY, TIME and CONSCIOUSNESS

RONALD SHARP

This book is © Copyright 2015 Ronald Sharp

Cover photograph of Ronald Sharp and glider 1957

Covers set up by Sandra Fogarty

GRAVITY TIME AND CONSCIOUSNESS

A NEW THEORY FROM FIRST PRINCIPLES WITH A SURPRISING RESULT

2015 Copyright © Ronald William Sharp BEM.

This essay is for those who are confused by advanced mathematics, the 'Big Bang' and the 'Expanding Universe'.

For Cosmologists who are not happy with current theory, the author would like to invite them to apply professional methods to investigate the direction that this theory of gravity based on observation, logical lateral thinking and forthright assertion proposes here.

The essay is an up-to-date summary without diagrams from my little book not yet completed;--

WHAT IS GRAVITY REALLY AND WHAT IS THE FOURTH DIMENSION ?

The book was commenced to be written down in 1992, bringing together and adding to thoughts collected from when this author discovered high school physics at age 12.

2.

At age 12 in 1942 I entered Kogarah Intermediate High School and was introduced to the subjects of Physics and Chemistry.

Sometime thereafter, a school friend, Adaver (Lou) Pedricks, said to me;
"Did you read about this scientist, Einstein, he has a theory called Relativity and only two people in the World can understand it".
Sir Arthur Eddington was said to be the second person. The news of his death in 1944 could have been the source of my friend's statement.

Me: What is Relativity?
Lou: It is about Time and the Fourth Dimension.
Me: What is the Fourth Dimension?
Lou: They think it's Time.
Me: But why is it called the Fourth Dimension?
Lou: Well, if you move a point to make a line, that's the First Dimension and if you move the line across to make an area, that's the Second Dimension and when you move the area up to make a solid, that's the Third Dimension.
Me: Oh! Then the Fourth Dimension is: _____

In the fifty years since then, between earning a living and other activities, I have thought about this, what is Gravity, Time, Space, but the source of information has been from mostly populist literature, radio, and TV.
During the late 1940's; subscribed to Sky and Telescope magazine and made a six-inch Newtonian telescope, grinding the mirror from Pyrex.

4.

In the early 1950's; designed, built and flew a ten-metre-span glider; applied unsuccessfully for a Draftsman's position at Mt.Stromlo Observatory in Canberra and applied for a Draftsman's job at the C.S.I.R.O. within the University of Sydney, enclosing the results of a vocational-aptitude test.

The interviewers looked me over and said, "Well, we can't give you the job because you don't have the qualifications, but were so taken with your I.Q. that we just wanted to see what you looked like".

During the early 1960's I was inspired by T.V. broadcasts, arranged by Professor H. Messel, in the Physics Department of Sydney University; of discussions on Cosmology; by Bondi, Gold, and Gamow, on such topics as The Steady State, The Big Bang and Hubble's Red-shift. Later in the 1960's Lady Suzi Jeans, widow of the physicist Sir James Jeans, played my Knox Grammar School, Opus 9, Pipe Organ.

A few years later, I worked jointly with the Hamburg Organ Builder, Rudolf von Beckerath, to build the new pipe organ in the Great Hall of Sydney University and following that, my own project, the creation of the Grand Pipe Organ for the Concert Hall of the Sydney Opera House.

I had not become aware that my intuitive reply about the fourth dimension which led immediately to a description of what Gravity could be had been put forward by anyone else.

Whenever I met people in the field of physics and astronomy I attempted to put forward this idea in the hope that it would offer a key to further investigation, but in most cases received responses ranging from indifference to derision.

Copyright © Ronald William Sharp 2015.

WHAT IS GRAVITY REALLY AND WHAT IS THE FOURTH DIMENSION

INTRODUCTION
This essay was to be the Chapter "Gravity" from memoirs yet to be written.

It is presented as my personal ideas on Cosmology, which have been considered from a Lay-observational point of view over fifty years, with the hope that it just might be a useful lead to further, more professional insight, as from what I have read, many of the problems discussed are said to be yet unsolved and subject to considerable theorizing.

If some of the ideas in this essay seem to be presented as established fact, it is because they are meant to represent my conclusions at the time and it is expected that response for further discussion will follow. Without conviction it is difficult to proceed in a field so dominated by established concepts.

There have been many theories that although seen at the time as possibly the answer to mysteries in Physics have later

6.

been superseded. Many of these theories have been proposed
by people of high repute.

I formed the idea that if these theories can be taken as
interim, this being a natural course of research and a quest
for understanding, then I may as well put in my bit, however
naive it may appear at first reading.

A STRONG INTUITION CAUSED THE BELIEF THAT

There was no such thing as a force of attraction anywhere in
the Universe.

The Hubble Red-shift did not denote that the Galaxies were
moving apart at ever increasing speeds of Earthly Miles-
Km./Hr with distance.

Special Relativity was flawed due to assumptions made on
the basis of earlier theories that had themselves been made
on assumptions and that the use of the word "Observer"
brought about difficulties.

Space was not a vacuum of nothing.

The Idea that the perceived force of Gravity was the result of
an inertial resistance to Acceleration came from my intuitive
description of the Fourth Dimension, to a high-school friend.

The belief that this same effect may also apply, but in
different forms, to Magnetism, Electricity and the Nuclear

Forces, was later prompted by learning of Einstein's quest for a Unified Theory.

THE CONTENT OF THIS ESSAY CAME ABOUT IN THREE STAGES

The initial idea of Gravity and the Fourth Dimension in the mid 1940's.

The renewal of interest on learning of the Hubble red-shift and the Steady-state-Big-bang debate in the early 1960's.

The desire in 1992, to write it all down, led to reading locally available literature in order to assess whether my ideas were compatible with professional progress.

FROM THE LITERATURE AVAILABLE TO ME, MANY THINGS JUST DON'T ADD UP

Special Relativity related to Observers and what they see, being taken not as optical illusion, but fact. Some of the popular explanations of Relativity seem to cause confusion and misunderstanding.

Simultaneity, claimed not to exist, (due to "c" the speed of Light).

Maxwell's Electro-magnetic "leap-frog" out of phase, yet in-phase.

8.

The belief that there can be forces of attraction.

The Theories that attractions are caused by appropriate particles.

The <u>entire</u> Universe created by a Big-Bang from a tiny point.

The simplistic mathematical idea that if the Universe is expanding, then going back in time leads to a tiny point or nothing.

The Michaelson-Morley Test said to prove the absense of "The Aether".

The operational-working-out of the Universe in natural, dynamic, three-dimensional Euclidean Space, converted to a complicated, manufactured, twisted-up, static, non-Euclidean Space.

How objects can accelerate in a neutral Inertial Frame, along lines of Gravitational-Space-Curvature, when we all observe and feel that "something" just HAS to accelerate "something-else".

Confused explanations of curved Space-time, by reference to two-dimensional beings on the surface a sphere. There can't BE two-dimensional beings. The manner of explanation, leads to the wrong interpretation of Curvature.
Thought experiments based on what just cannot BE. (Often give false leads).

Discussion about whether the Universe is open or closed, dependent on its density related to the Hubble constant of expansion, which assumes that the Galaxies are flying apart at ever increasing speeds and that the whole Universe is a sphere.

The rubbishing of Laplace for the elementary observation that everything evolves from moment to moment according to the Working out of the Way of the Universe. (The Real Laws of Physics).

Explanations, of the Fourth Dimension as Space-Time-diagrams called World-lines, causing a diversion from seeing the real Fourth Dimension.

A straight line IS a straight line in Space. If it curves, it does so for a reason that is explainable in Euclidean Space. There is no need to create a theoretical Space instead of persuing the realistic cause as we would see it.

Please read the Essay sequentially, in order to obtain a feeling for the way-of-thinking that has evolved the views expressed.

Don't peek ahead.

Copyright © Ronald William Sharp 2015

(End of extract)

10.

PART 1.

This initial test to try to understand Gravity and Acceleration may be done using say, a bottle, full or empty, with a handle for a finger-hold and a very light-weight toy wheeled cart on which the bottle can rest horizontally.

With sufficient experience in mechanical work and an elementary knowledge of physics, it is possible to dispense with this equipment and make the test as a thought experiment…. Einstein made thought experiments.

Hold the bottle steadily by the finger-hold. The force felt by the finger is the weight of the bottle at the earth's surface. This is also its mass. The mass doesn't change because that is what it is, a collection of atoms and molecules etc. but the weight changes according to Gravity, at locations such as the Moon and also by Acceleration.

The acceleration due to gravity at the Earth's surface is about ten metres per-second per-second, which means that if one lets a mass drop, the speed of fall will increase by ten metres per second, every second.

If a person falls out of an aircraft without a parachute, then that person will reach a speed of fall of around two-hundred Km. per <u>hour</u>. Air resistance restricts further increase in speed of fall… From Space with no air resistance the speed of fall could reach over ten Km. per <u>second</u>.

The weight of the bottle and indeed also of the person holding it is called "one g", that is, one unit of gravity at this place on the Earth's surface.

12.

If the holder were to be able to accelerate the bottle upwards at ten M/s/s, the force on the finger would be twice that when at rest, or "two-g", one g" at rest plus "one g".

With sufficient distance under the bottle, if one were to step off a high ledge, then the force on the finger would reduce to zero. "zero g". The holder would fall with the bottle and they would be accelerating downwards at ten M/s/s.

Back on the ground, accelerating the arm downwards at twenty M/s/s *equivalent* to "two g", the bottle would invert, trailing the finger as it were pulled downwards with a force on the finger of only "one g". This is because although the acceleration downwards was at the *rate* of twenty M/s/s, the *force* would be that of only "one g", because the acceleration of Gravity, "one g", must be deducted from the actual acceleration of "two g".

Placing the bottle on the toy cart and then pulling it along a table or pathway at an acceleration of ten M/s/s, the force on the finger would be the same as when holding it upright at rest "one g". Pulling the bottle at twice that rate of acceleration, the force on the finger would be twice that at rest, "two g".

With these observations thoroughly absorbed and well understood, it must be clear that the forces and the accelerations from which they emerge are all of the same nature. The force on the finger when pulling the bottle horizontally is the same as the force on the finger when holding the bottle at rest. ..... *See Note 1. P.33.*

Although Einstein said that the forces of Gravity and Acceleration were *Equivalent,* it is clear, without previous reading-influence on the subject and <u>with only what has been observed here</u>, that Gravity and Acceleration are not *equivalent*, but equal, they must be one and the same phenomenon:.....Has it not just been observed?
Once this has been accepted, progress may be made in simple logical steps.

The first observation to be made from the above is that the surface of the Earth (joined to one's finger) on which the experiments were made, is accelerating upwards at "one g" and as the surface at the antipodes would be doing the same thing, the Earth must be expanding, with everything else in our vision and experience expanding with it.

Therefore, the conclusion of Part 1. is that Gravity *appears* to be due to the accelerated expansion of massive bodies.

For any valid contemplation of our "World" or "Universe" or the "Infinity of Space", it is necessary to keep in mind just how extensive it all IS.
Start with a roughly scaled model of our local area of space.

A scale model of the Moon the size of a grain of fine sugar.
The length of a finger away, the Earth is the size of a grain of sand.
Forty metres away, our Sun is the size of a Soccer Ball.
At this same scale of size and distance, how far away is the nearest Star, like our Sun?
It is the distance from the east coast of Australia to Central America.

14.

Just imagine we are out in Space looking "down" at this almost unbelievable reality!

But inside the Earth, represented by the sand grain, is an immense number of atoms and elementary particles that are of an equally unbelievably small size. They are described as having mostly "empty" space separating them, therefore they may look similar to the distribution of Stars in our scale model!

PART 2.

How can it *be*, that the logical conclusion of Part 1. is really true?
All the matter in the Universe surely cannot be expanding exponentially whilst keeping the same proportions and fundamental distances apart.
How can it just expand forever? The speed of light would be reached in a short time.

<u>There are at least two observations that refute the conclusion of Part 1</u>.

Massive Bodies could not ORBIT.
Whilst expanding, they would just travel in straight lines past each other.

If there could be a tower with a height equal to Earth radius, the gravity due to expansion, at the top of the tower, would be twice that on the surface, twenty M/s/s., because it would be expanding at twice the rate at the surface.
Due to its reduction at the square root of the distance, Newtonian Gravity would be one quarter that on the surface, one eighth of that by expansion. (This may not apply exactly so close to the Earth's surface).
It is therefore clear that the conclusion of Part 1. is <u>not correct</u>, even though it seems to be the only answer to the simple logical experiments.

<u>How may this quandary be resolved?</u>
If the Earth is expanding, then into WHAT is it expanding?

16.

SPACE?

What is Space?   Is it Something?   Is it Nothing?   Is it Infinite, extending forever?

If it is <u>absolutely nothing</u> then where did all the mass bodies come from? Surely nothing can come out of nothing?... Some say that the Universe of matter exploded from a concentration of matter the size of a pin head!!??

So Space must be <u>something</u>, if so, what? Many have called it the Ether.
Surely it must be infinite, extending forever. If it were not, then where would be a boundary between something and absolutely nothing? How would the currently accepted theories of gravity keep the something and its contents together in such an unreal situation?

This essay is a summary of the book mentioned above, in which a number of elementary observations are perused. The whole theory is based on simple mechanical logic by which we all live. It starts from the <u>obvious</u> place, <u>experiencing gravity and acceleration</u>. Many theories take over from others using mathematics and are obliged to maintain the formalities of the system, even if common sense shows doubt.

Here are stated briefly some commonsense assumptions that will allow proceeding with our puzzle.

<u>Space is infinite</u>, it extends forever, it always has and always will. We all should avoid spending our lives arguing about it

and get on with more relevant issues. This surely cannot be proved but is the most reasonable conclusion in order to move on.

In this regard, the CMB COBE fuzzy images are most likely to be of distant universes or groups of galaxies and not remnants of radiation left over from the so-called Big Bang. There are many bangs, small, medium and big, occurring throughout infinite time.... *See Note 2. P.33.*

Space is something, not a vacuum of absolutely nothing. Let's call it the Aether, it is not an anaesthetic solvent called Ether. Space can flow and is the ultimate basis of everything. As mentioned above, we shouldn't waste time debating it just now as it is necessary to move on.

Space has the characteristic that is the next most fundamental basis of everything, a... Response to ACCELERATION. (The rate of increase in speed).

When Space is disturbed, as it may be by waves, or the acceleration of a Mass Body, it resists movement in non-linear proportion to the speed and acceleration *within* a wave. Anyone who has helped their mother mix custard powder will know that at a certain consistency, or viscosity, stirring with the spoon slowly produces little resistance. The faster one stirs, the greater the resistance, until with a very quick movement it is impossible to move the spoon at all until the force that moved it is relaxed. The thick uncooked custard can flow and be poured, but cannot be forced to do so at a rate higher than its fundamental nature will allow......*See Note 3. P. 33.*

In this description, the custard must flow around the width of the spoon to get out of its way. This means that it must move faster than the spoon is moving because it is directed sideways compared with the direction of the spoon's movement. This multiplies the effect of the forward acceleration of the spoon by lateral motion of the medium.

If a wave is graphed on paper with height for amplitude and horizontally for progression, two characteristics will be observed. The slope of the line changing vertically represents the speed within the wave and the radii at its upper and lower reversals represent the internal acceleration. The steeper the slope, the higher the speed and acceleration *within* the wave. Call this the **Wave-front-slope.**

So the Universe is made of Banana Custard after all....
Don't laugh too much yet._____All will be revealed.

These modeled characteristics of Space allow further progress towards a new theory.
How may the observation that gravity must be mass expanding into space be explained, without its actually doing so?
Reverse the idea and see it as space contracting into mass or space flowing with acceleration into mass. This would eliminate the quandary from Part 1. and still conform to the observed effects.

The idea that gravity results from space flowing with acceleration into mass may explain much of what we observe in the working out of the way of the universe, but it must be remembered that mass and gravity are inextricably linked, so therefore gravity must act on the smallest particle that can be called mass-matter. Inflow must therefore seek out the smallest mass particle in any solid body, regardless of the total mass. As gravity exists while matter exists, where would all the inflow of space-aether congregate, or does it form the continuing state of the matter?.....*See Note 4. P.34.*

Inflow may at last explain how objects are gravitationally "attracted" because inflow of space into each would ultimately allow them to be "pushed" together regardless of distance, because of the diversion or dilution of space flow between them.
However at <u>great distances</u> between them gravity may weaken or even reduce to zero due to the inflow of infinite space making up for the dilution mentioned above.

<u>This might be interpreted as the expansion of distance</u> between all the matter within space due to the *apparent* exponential weakening of gravitational "attraction". This however is not a valid assumption as careful study of the previous paragraph will confirm. It may also be the reason for Einstein's "Cosmological Constant", meant to oppose this supposed expansion.

Inertia works due to the acceleration *within* waves of space, mentioned in paragraphs earlier. Acceleration and motion of a mass body pushing through space would increase the wave-front slope of the produced wave in space, showing up as inertia in similar fashion to the description of the spoon in custard. Inertia is the resistance to acceleration. <u>If the</u> <u>fundamental nature of space-aether is a response to</u>

acceleration or an exponential response to movement, then
this is a description of how Inertia works.

If gravity is the inflow of space-aether into mass then the
space-aether flowing vertically past an object into the surface
of the Earth will exert a force on that object, created as above
between that object and the accelerating space-aether. This
would be what we feel as gravity.

Acceleration and inertia may also explain the nature of time
and in the story of the Twins paradox, the reason why the
twin who goes away into space is younger on return.

**Time could be the sequential accumulation of the
smallest increments of *change* in the smallest particles
that can be called mass-matter.** Acceleration could slow
this rate of change, therefore slowing time. The twin who
goes away accelerates four times, twice speeding up and
twice slowing down, in order to arrive back, therefore has
slowed time, becoming younger than the twin who stayed
behind. Time could flow more slowly in aircraft due to the
accelerations of flight, such as take-off, air turbulence and
landing, compared to those left stationary on the ground.

Would not this infer that time flows more slowly on the
Earth's surface than in space, as the acceleration due to
Gravity is stronger, nearer to the mass body?

Gravitational Inflow of space would be almost vertical to the
Earth's surface (objects fall vertically) and as the apparatus
of Michaelson-Morleys' experiment would be positioned
only the thickness of a sheet of paper from the surface of a
model Earth having a diameter equal to the height of the

Empire State Building, they could not have detected a flow of Aether past the apparatus and the Earth _horizontally_. They did not therefore demonstrate that there was no space-aether.

The speed of light is described as about 300,000 Km per second.
100 Km per hour on our highways is only $1/11,000,000^{th}$ that of light speed.
30,000 Km per hour for a satellite is only $1/37,000^{th}$ that of light speed.
Like thick uncooked custard, space-aether reacts to acceleration exponentially thereby allowing a body with a very slow speed to move through it with almost no resistance because the wave-front-slope is so low and insignificant compared to what it becomes nearer to the speed of light. However, acceleration changes this.
This must be how massive bodies would move through a something space apparently without friction, if not influenced by the gravity of other matter.

A graph can be described to help explain this.

Draw a horizontal line 20 cm. long representing _linearly_ from left to right, the speed of light from zero to 300,000 Km per second.

At the right end draw a 10cm. vertical line representing acceleration, wave-front-slope within a wave in space.

From the left end, just touching the base line draw an exponential line almost parallel to it for about 5 cm. then curving up ever more steeply so that it is parallel to the vertical line when it meets this line for 1 cm. at the top.

This curved line then represents the resistance to speed in Space from almost nothing at the left end to being unable to pass the vertical line at the speed of light. Note that this is similar to the action of stirring the custard in our analogous experiment.

From the above, note that when moving at low speeds like 100 Km/hour on Earth or 30,000 Km/hour as a satellite, there is in practical terms, negligible resistance to motion.
The distance from the left end of the graph for 100K/hour is 1/55,000 of a millimeter and for 30,000 K/hour is 1/183 of a millimeter.
It is no wonder that there is almost no resistance to this relatively slow movement through Space. A body might travel with momentum at almost uniform speed once started, but will encounter a stopping force the more it is accelerated until it cannot exceed the speed of light.

This is why Newton's gravity and General Relativity gravity are almost the same at speeds and accelerations relative to our World.

However, neither explains what Gravity really IS. They are only mathematical constructs that allow predictions to be made.

The larger the mass to be accelerated towards the right of the graph, the steeper from the base line must the curve start, because the greater mass will create a greater wave-front-slope in both space and itself. The vertical line at the right may need to be longer.

Although it seems that a mass body may move slowly compared to the speed of light with uniform motion through the space-aether, it is also under the action of the inflow of that same aether to create its own gravity. This would result in a re-direction of the in-flow vector.

The Earth's daily rotation slows and satellites of the Earth and Sun slow down. Perhaps the foregoing thoughts explain this. They are however under the acceleration of their parent bodies and this must be a slowing force, due to the effective increase in speed relative to space-aether with this acceleration towards the Parent body.

It would seem that acceleration in our low speed environment produces inertia, whilst a steady low speed produces little impediment to progress.

In perusing, over a long life, descriptions of the universe of stars and galaxies and marveling at the pictures from optical and radio telescopes and lately from the Hubble telescope, one observes that there can be no part of our universe that is free from the influence of the gravity of other matter-mass. Therefore there can be no neutral area in all Space that does not contain the gravitational influence of mass bodies.

Many, including this author, who could not understand the mathematics of General Relativity have wondered how a mass body could travel through a curved corridor of "space-time" without needing an accelerating force to change its direction. This did not seem to match the common sense of ordinary experience.

Earth gravity accelerates an orbiting body to change course continuously by vector. The orbiting body is moving forward with almost no resistance from Space due its low wave-front slope but is being continuously accelerated towards the parent body by Gravity.

The "inertial frame of reference" and "space-time" therefore seem to be only mathematical constructs in the absence of a description of the reality of gravity.

Regardless of the advantages of the inflow theory, it is difficult, as it was for Part 1. to accept that space-aether flows into matter forever and this intuition leads to a more acceptable theory in known mechanical terms that are the basis of a logical progression of ideas and <u>may describe somewhat equally the results of space-inflow.</u>

PART 3.

How may the difficulties of infinite inflow of space-aether into matter be described in a way that is more acceptable to our view of how our World works?

Could a great number of small increments of accelerated inflow replace a smoothly accelerating inflow of Space?
As with the Inflow theory this must connect with or start as a reaction between space and the smallest particle called Matter.
Would this be termed a <u>Quantum state of Gravity</u>?

How could small increments of an inward accelerating Space be described?

Gravity is a negative (backward-accelerating) sawtooth longitudinal wave propagating outwards into Space from the smallest particle that can be described as matter/mass.

The high-slope part of the sawtooth wave (due to the momentary increased acceleration in the Space medium back towards the source), creates the <u>Force of Gravity</u> imposed upon another body or particle of matter. It is a spherical wave that propagates (presumably) at inverse square. (The wave must run-down eventually and exponentially in energy over time and distance as does everything, including light).

<u>For this to happen</u>, Space has to be <u>something</u>, with an all-pervading pressure.

Without this pressure, a negative-biased wave, (a Graviton?) cannot propagate from its source, unless the source is hollow relative to Space.
The smallest particle that can be called Matter must <u>pulse</u>, or <u>vibrate in an exponential sawtooth manner</u>, with high acceleration inwards and low acceleration outwards.

To make up to this pulsing, <u>Space flows momentarily</u> towards (due to its pressure) and away from the mass particle, so creating a wave that propagates away from the mass particle.

The Gravity we feel is a smooth accumulation of the many individual (Quantum) accelerations of Space on the smallest particles that can be called Matter.

This pulsing may relate in frequency to an explanation of Wave-Particle Duality expounded in the book referred-to in this essay.

Gravity works on all Matter, therefore an explanation must be related to Matter, what we are and what we see, and that is a Mechanical explanation.
For a mechanical explanation, Space has to be Something, not Nothing.

The pressure of Infinite Space is within us just as the pressure in a deep ocean is within the fish and organisms that live down in the depths. They don't feel it and neither do we. There is no way to prove this so we should accept it as very possible and move on.

The Casimer Effect is associated with an experiment to discover the pressure of Space and other effects. It used two metallic plates placed very close together. The basic principal was to exclude space from this close gap so that the pressure on the outer surfaces of the plates would push them together. A small force was measured that was taken to be the result of fluctuations in Space or electro-magnetic effects.

As Gravity acts from the smallest mass particle and Space invades everything, then the pressure would be all through the two plates and Gravity would act between them.

Therefore, the Force measured would have been from the Gravity between them as described in this essay plus a very tiny force from space-pressure-effects between the outer surfaces and the gap. But not the total real pressure because this would already exist between all the elementary particles, within the plates.

Please refer back to the scale model of the local area of our Universe in Part 1.

From this picture, that the scale distance of the nearest star to our Sun the size of a soccer ball, is the distance from Australia to America and the nearest next star is that far away again, we may go on and on until we have accounted-for.all the stars in our milky way galaxy.

Our Milky Way galaxy of two-hundred billion stars is over one hundred-thousand light years in diameter. One light year is the distance (relative to our standards) that light travels at a speed of three-hundred-thousand Km per second in the time (in our standards) it takes the Earth to orbit our Sun (relative to the infinity of Space).

28.

From this view of our galaxy, the next major resident of Space, the Andromeda galaxy, just like ours but twice its diameter is two and a half million light years away and then on and on again to about one-hundred billion galaxies in our visible universe.

Current thinking from Hubble telescope images infers that the youngest (earliest) galaxy so far seen <u>was</u> fourteen billion light years away... Fourteen billion <u>years ago!!</u>

There is a mystery that is recalled by the reference to the Hubble telescope. Edwin Hubble discovered that there is a connection between the reddening of light from far away galaxies and their distance from us. The greater the distance, the redder the light, on a linear scale.

Much discussion has ensued as to whether this is due to the Doppler-effect, the <u>acceptance</u> of which leads to the theory that the galaxies are moving away from us, (like the lowering pitch of a siren after it has passed by). The further away they are, the faster they are moving. This has led to theories of expansion of the universe, the big bang and the cosmic microwave background. By using the formula for expansion in reverse, some have theorized that the whole universe exploded from a concentration the size of a pin head and that they can describe the state of this explosive mass at less than (as quoted), "one <u>earthly second</u> from the beginning".!!??

The whole edifice is built on the assumption that light does not vary in its energy and wavelength no matter how many billions of years in time, distance and gravitational

acceleration through which it has propagated:......Is light the only thing in all creation that does not run down?
Do all things based on electro-magnetism never change or run down?
Do elementary particles run down? In the LHC they do when viewed as smashed bits of Protons. Perhaps these are only potential particles.

Others including the great Paul Dirac have suggested that light gets tired so that the wavelength increases towards the red end of the spectrum. It could be assumed that all wavelengths down from gamma rays to long radio waves and indeed also down to the longest of all, radiation from other universes distant from our own run down in energy, thereby increasing their wavelength. If this is so, the theory that the universe is expanding is not correct and by inference the edifice built upon it would not be valid. Which theory is correct?

Referring back to the paragraph in Part 2. describing the speed and acceleration *within* a wave in space, it would seem that a lessening of the acceleration (energy), (during time and distance traveled), at the change of polarity results in an increase in wavelength towards a reddening of the light. This can be sketched as a graph.

Although light is described as an electromagnetic transverse wave it is nevertheless propagating in space, which must be the "ultimate basis of everything"..... It is difficult to understand how this wave which is accelerating four times a cycle_trillions of cycles a second between an electric action and a magnetic action in a leapfrog sequence can go on forever without running down like everything else in existence.

It is stated that the speed of light is "Constant in a Vacuum". But there is no Vacuum. So how can it be constant? Was it meant, the "Vacuum of Space" or one in a container?

Does not everything that runs down accelerate the running down the longer it goes on? Would not this apply to light? If this is taken to be a Doppler shift it is no wonder it is believed that the Universe is expanding exponentially. How can Space expand when it is infinite, extending forever?

What if, (a very useful phrase), the red-shift were to be taken as light running down in energy exponentially after a certain time and distance of stability and then applied to fresh theories of the Universe as if there were no previous theories from which connecting mathematics could influence a result?

At the scale of our model, where the Earth is a grain of sand,.... where in reality, is the true basis of the rules for the working out of the way of this universe? Is it our Earth or is it the whole of Space containing a multitude of Universes?

Within Humanity, great minds have worked out many of the circumstances that have evolved over millions of years, (revolutions of the Earth around the Sun relative to space) but everything we experience here is the result of the evolution (in local terms) of bits of Matter that have accumulated by the action of gravity after floating around in

Space. Are these bits of Matter really what is being described as Dark-Matter not yet clumped together enough by Gravity to initiate fusion and light? The creation of Stars.

Think about reality. There is no colour in light, only a frequency or an acceleration that is interpreted by our brain. This interpretation is accompanied by **the process of consciousness** which itself doesn't exist as a thing. With no process of one action after another through Time, there is no consciousness (a process akin to the scientific method). Observation, input, comparison with memory, make hypothesis, compare with next input and so on as a process. This process has a frequency yet to be researched but is most likely to vary according to the living organism, from say trees, the slowest, to the tiniest of insects, the fastest or even down to microbes.

Without the lens of our eye, camera or telescope to bring rays to a focus, there is only a fog of the radiation we call light dancing around us. Visual reality is a diffusion of energy waves, not an image and without us to interpret it, not even a grey fog, as we might say "just blackness".
Dwell on this.

Please re-read the last two paragraphs.

With this in mind, how important is the Earth and its contents in, "the working out from moment to moment of the way of the Universe", a phrase that may solve a great puzzle that has mystified people for thousands of years.
It is therefore necessary in our perusing the nature of Space and its Gravity to accept that we are not the centre of the

Universe and to look at all aspects of our existence from the wider perspective.

This essay is not about electromagnetism, (whatever that is likely to be), but Matter and its Gravity as we know it. It is all mechanical as we see it and feel it. We, the reader and I, started this from the actual experience of a piece of Matter, a Mass and its association with what we know as the effect of Gravity.

The search continued with ordinary everyday logic from one step to the next, with only high-school physics, popular reading and experience with the mechanical world around us.

The Universe doesn't know about mathematics, <u>it just goes about its own process.</u>

This theory of Gravity and the further steps in understanding from here on must take in the whole picture, using all the amazing findings of Science and its adherents right or wrong from our past.   A task that may be outside the qualifications of this author.

Note 1.

The value of acceleration is not strictly true. Newton's gravity requires combining that of the Earth with that of the bottle. When pulling the bottle horizontally, the combined gravity is vectored, that is angled, therefore for a certain force on the finger, the acceleration would be different from what it would be if Earth gravity were not present. However, this observation does not affect the outcome in principle of these experiments.

Note 2.

CMB refers to the Cosmic Microwave Background radiation, presumed to be left over from the "big bang". COBE refers to the explorer satellite that photographed these fuzzy images.

Note 3.

When this author was growing up in the great depression, he would help his widowed mother mix custard powder for the usual banana-custard pudding.
As the water or milk had to be added slowly, at a certain consistency it was discovered that if the mix was stirred too quickly the spoon would come to a dead stop. No amount of force would allow it to be moved on.
This curiosity to experiment has lasted into old age and resulted in a number of achievements. It would be satisfying to find that this childhood observation had led to the results expounded in this essay.

34.

Note 4.

Without pondering too deeply at this stage of the essay how matter was created, could it be that the inflow of space-aether maintains matter in its continuing form? Or could it be that matter is <u>continuously created</u> by space-aether inflow? E=m.c-squared shows how great the amount of energy (space-aether?) is needed to make a unit of mass....... but how much energy is needed to maintain the matter?
What happens when a mass body gets so large that space inflow cannot reach the particles near and at the centre? Would not the central particles revert to energy? This energy could be released dependent on the size and shape of the mass body.

It could be anything from Jets to Super-novae or even Bangs.

Perhaps Matter is maintained by its access to space Pressure.

A critical mass of unstable uranium will explode if spherical, but will that same mass explode if drawn into a wire one mm. in diameter with only half a mm. to the centre?

Note 5.

One aspect of Light is resonant energy released from atoms when electrons accelerate back to their usual orbit-shells after being raised to new higher-level orbit-shells by the input of energy including light.

This output energy has a colour-frequency identified with the particular atom due to the harmonic partials represented by the spacing of the orbit-shells.

Electron orbit-shells may be harmonic standing-wave patterns caused by reflections from the atom of its gravity quantum-inflow accelerations.

An analogy of this can be observed when blowing harder, with a higher pressure-energy, wind instruments including organ flue pipes. The harmonic partials which are standing waves within the resonator, jump up to their next level, creating an octave higher sound. When the normal pressure is restored, the partials fall back to their original levels.

36.

TIME..... <u>Everything</u> exists forward through Time.

Time cannot go backwards.

Time is clearly the accumulation, one thing after another, of Something.

This must be the sequential accumulation of the smallest increments of <u>change</u> in the <u>smallest particles of Matter.</u>

Parts of a vibration or resonance or whatever it is. Is it Planck's length or the Higgs Boson?

Smooth <u>Gravity</u> is the accumulation of many small accelerations of space towards the <u>smallest particles of matter.</u>

Because they are both produced by the same actions of the same particles, <u>Gravity and Time must be decisively connected.</u>

Call them <u>Gravity-Time.</u> (Not Space-Time).

Copyright © Ronald William Sharp 2015.

A THEORY of WAVE-PARTICLE-DUALITY

The electro-magnetic spectrum reaches up to the wavelength of Gamma waves from very long radio waves and lower.

Einstein's exposition of the photo-electric effect shows that for waves of light to knock an electron from a piece of metal or to seriously disrupt its wave-form, it is not the quantity or magnitude of light waves but their energy, their internal *acceleration* that does the necessary work.

A graph may be drawn to represent a wave. (Not to scale)

The base-line represents Time or Speed of propagation, either will do to begin.
The height of the wave, say 20mm represents "h", Planck's constant.
Draw a sine wave with say 40mm between wave-peaks.

The slope of the lines changing vertically represents the speed within the waves and the radii at the upper and lower reversals represent the internal acceleration. The steeper the slope the higher the speed and acceleration *within* the waves. The frequency also rises because the height of the graph representing the amplitude of the waves does not change. E = "h" x Frequency.

If one moves a pencil, held flat against the paper and at right-angles to the base-line, from left to right, through Time, one sees how the *wave-front-slope-for-speed* and the *acceleration-for-energy* are represented by the graph.

38.

If the base-line now represents speed of propagation, a <u>wave-front-slope</u> of forty-five degrees should show that the speed within the wave is the same as the speed of propagation.

Would this mean that at the propagation speed of light, the speed within a light wave is more or less than that or, is the true speed of light really that *within* the wave? Is the height of the wave really "h" Planck's constant in Einstein's photo-electric-effect?

If increasing frequency becomes matter when wave-front-slope reaches the critical conversion acceleration angle and if the real speed of light is that *within* the wave, it may explain why matter and high intensity waves propagate at less than light-speed.

The distance between the wave-crests represents the wavelength and at a specified speed of propagation, the frequency.
What would be the wave-front-slope-angle for say a Gamma wave?

With the above thoughts as a basis for further progress in simple logical terms, a theory of Wave-Particle-Duality and also of Matter-Creation comes to mind.

It is popularly known that Ultra-Violet light causes sunburn and skin cancer.
Next upward in frequency, X-rays act like little bullets to create images of bones.
Next up, Gamma-rays are considered lethal due to their resemblance to real bullets.

What is next in increasing frequency?
Beta Waves?   Alpha Waves?   Potential Matter?   Real Matter?

Each of the above items increases in frequency; and for a fixed amplitude, say Planck's constant, also increases in Wave-Front-Slope, that is, the *speed* and *acceleration within* the wave.   Their Inertia increases with frequency as does their apparent Mass, due to the *increased acceleration*. Inertia equates with Mass.

However, Gravity initiates from the smallest Mass particle, whereas Light is said to initiates from the Electrons of Atoms. *See Note 5. P.35.*

In Audio engineering, it is considered impossible to reproduce or create a square wave. The wave-front-slope would be vertical, representing an infinite speed and acceleration within the wave. In reality, the wave produces "ringing" as a protest against forcing the system to over-reach the possible. The "ringing" is shown as spikes on the wave and results in the distortion of sound reproduction.

A Tsunami wave is a Soliton (a raising of a volume of water due to uplift in the ocean floor) and it starts moving with an almost undetectable wave-front-slope but resulting in a massive wave-front-slope at the shore.   Although the new volume of water propagates outward in a circular wave, enough of it reaches the locations of massive damage.

In contrast, Wind-developed Ocean Waves travel without the water actually traveling with them, but when the ocean floor rises near the shore, the wave-front-slope increases until at the shore line it <u>exceeds</u> the <u>critical wave-front-slope</u> and topples over, breaks, creating a Soliton (potential Mass) that propagates up the beach. (Wave changes to moving potential solid).

What happens when, in Space, waves caused by explosions or Standing Waves developed from two or more propagating waves exceed the critical wave-front-slope and distort, turning into solitons of temporary matter or in extreme cases Permanent Matter. Statements from Science that the Universe is full of particles of matter that appear out of nothing only to disappear almost immediately are most likely the result of the foregoing.

In comparing the frequency of radiation that as above could be <u>the creation of matter</u> and the frequency of the smallest particle that could be called matter in relation to the source of Gravity, is it possible that that they are related or indeed the same?

It is also assumed that Time is the accumulation of the smallest increments of *change* in the smallest particle that can be called matter.

Could this mean that Gravity, Time, the creation of Matter and the maintenance of Matter are all related in frequency and have the same source. In the book from which this essay is a summary, Space-time is called Gravity-time.

From Page 3.

Me: What is Relativity?
Lou: It is about Time and the Fourth Dimension.
Me: What is the Fourth Dimension?
Lou: They think it's Time.
Me: But why is it called the Fourth Dimension?

Lou: Well, if you move a point to make a line, that's the First Dimension and if you move the line across to make an area, that's the Second Dimension and when you move the area up to make a solid, that's the Third Dimension.

Me: Oh! Then the Fourth Dimension is:_____

"The expansion of the solid in all directions outwards into Space."

"If the expansion ceases, then the solid just reverts to its state of being in the third dimension."

The fourth dimension is therefore a <u>Process</u>. Not a thing.

Would not this relate to the expansion of matter from Part 1 and from there to the Inflow of Space from Part 2, Thence to the Quantum inflow of Part 3, and on to the connection of Gravity and Time?

Copyright © Ronald William Sharp 2015.

*

*

*

www.ingramcontent.com/pod-product-compliance
Lightning Source LLC
Chambersburg PA
CBHW070230210526
45168CB00019B/1336